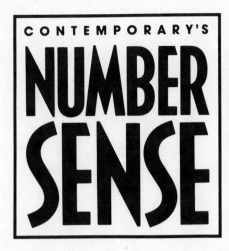

CONTEMPORARY'S

NUMBER SENSE

Discovering Basic Math Concepts

The Meaning of Fractions

Allan D. Suter

Project Editors
Kathy Osmus
Caren Van Slyke

CONTEMPORARY BOOKS

a division of NTC/CONTEMPORARY PUBLISHING GROUP
Lincolnwood, Illinois USA

ISBN: 0-8092-4226-5

Published by Contemporary Books,
a division of NTC/Contemporary Publishing Group, Inc.,
4255 West Touhy Avenue,
Lincolnwood (Chicago), Illinois 60646-1975 U.S.A.

8 9 0 C(K) 20 19 18 17

Editorial Director	*Cover Design*
Caren Van Slyke	Lois Koehler
Editorial	*Illustrator*
Seija Suter	Ophelia M. Chambliss-Jones
Sarah Conroy	
Ellen Frechette	*Art & Production*
Steve Miller	Andrea Haracz
Robin O'Connor	
Lynn McEwan	*Typography*
Lisa Dillman	Impressions, Inc.
	Madison, Wisconsin

Editorial Production Manager
Norma Fioretti

Production Editor
Craig Bolt

Cover photo © C. C. Cain Photography. HYDROX Creme
Filled Chocolate Sandwich cookies is a registered
trademark of Sunshine Biscuits, Inc. Reproduced with
permission.

Dedicated to our friend, Pat Reid

Contents

6 Fractions and Decimals

7 Life-Skills Math

Denominators

The bottom number of a fraction—the **denominator**—tells how many **equal** parts a whole object is divided into.	$\dfrac{1}{4}$ ◀── denominator

Below are drawings of whole objects that are divided into equal parts. Write the fraction that is shaded in each drawing.

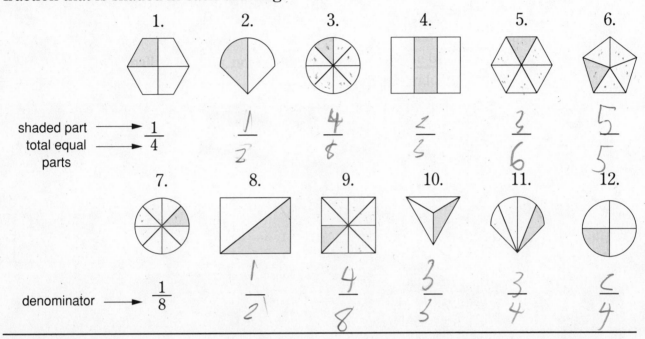

shaded part ──▶ $\dfrac{1}{4}$
total equal parts

2. $\dfrac{1}{2}$ 3. $\dfrac{4}{8}$ 4. $\dfrac{2}{3}$ 5. $\dfrac{2}{6}$ 6. $\dfrac{5}{5}$

denominator ──▶ $\dfrac{1}{8}$

8. $\dfrac{1}{2}$ 9. $\dfrac{4}{8}$ 10. $\dfrac{3}{3}$ 11. $\dfrac{3}{4}$ 12. $\dfrac{2}{4}$

Shade **only one** of the equal parts in each drawing. Write a fraction for the shaded part of each object.

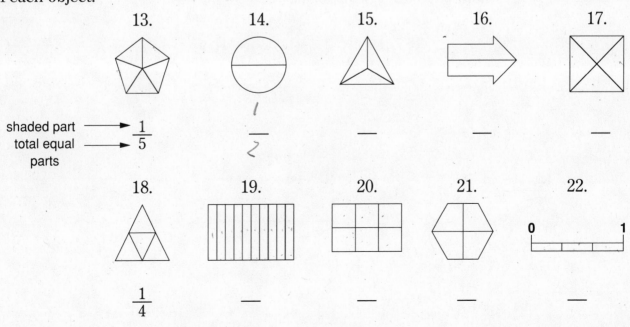

shaded part ──▶ $\dfrac{1}{5}$
total equal parts

14. $\dfrac{1}{2}$ 15. — 16. — 17. —

18. $\dfrac{1}{4}$ 19. — 20. — 21. — 22. —

Numerators

The top number of the fraction—the **numerator**—tells how many equal parts are used in the whole object.

$$\frac{1}{4} \longleftarrow \text{numerator}$$

Below are drawings of whole objects that are divided into equal parts. Write the fraction that is shaded in each drawing.

1.

fraction shaded $\longrightarrow$ $\frac{3}{4}$

2.

—

3.

—

4.

—

5.

numerator $\longrightarrow$ $\frac{2}{4}$

6.
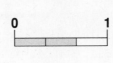
—

7.

—

8.
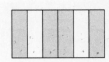
—

9.

—

10.

—

11.

—

12.

—

13.
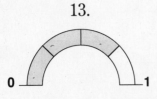
—

14.
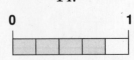
—

15.
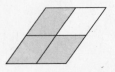
—

Fractions with Numerators of 1

Match each drawing with the correct fraction.

1. $\left(\dfrac{1}{5}\right)$　2. $\left(\dfrac{1}{8}\right)$　3. $\left(\dfrac{1}{9}\right)$

D

___letter___　___letter___　___letter___

4. $\left(\dfrac{1}{7}\right)$　5. $\left(\dfrac{1}{10}\right)$　6. $\left(\dfrac{1}{3}\right)$

___letter___　___letter___　___letter___

7. $\left(\dfrac{1}{4}\right)$　8. $\left(\dfrac{1}{6}\right)$　9. $\left(\dfrac{1}{2}\right)$

___letter___　___letter___　___letter___

A 　B

C　D　E

F　G　H

I

Draw and shade in each of the fractions below using the rectangles.

10. $\dfrac{1}{5}$

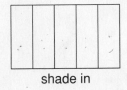

shade in

11. $\dfrac{1}{6}$

12. $\dfrac{1}{3}$

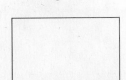

13. $\dfrac{1}{7}$

14. $\dfrac{1}{2}$

15. $\dfrac{1}{8}$

16. $\dfrac{1}{4}$

17. $\dfrac{1}{9}$

Match the Fractions

A fraction with a numerator smaller than its denominator is a **proper fraction.**

Match each drawing with the correct proper fraction.

1. $\frac{2}{3}$ 2. $\frac{3}{6}$

A
_____   _____
letter   letter

3. $\frac{1}{4}$ 4. $\frac{5}{6}$

_____   _____
letter   letter

5. $\frac{3}{5}$ 6. $\frac{2}{4}$

_____   _____
letter   letter

7. $\frac{5}{8}$ 8. $\frac{3}{4}$

_____   _____
letter   letter

9. $\frac{1}{3}$ 10. $\frac{1}{2}$

_____   _____
letter   letter

A   B   C

D   E   F

G   H

I   J

Draw and shade in each of the fractions below using the rectangles.

11. $\frac{2}{5}$ 12. $\frac{3}{4}$ 13. $\frac{1}{6}$ 14. $\frac{3}{5}$

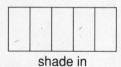 

shade in

15. $\frac{2}{3}$ 16. $\frac{3}{8}$ 17. $\frac{1}{2}$ 18. $\frac{3}{7}$

 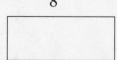  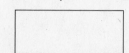

Fractions Larger Than 1

A fraction with a numerator larger than its denominator is an **improper fraction.**

Count the shaded parts and complete the number sentence for each drawing.

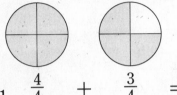

1. $\dfrac{4}{4}$ $+$ $\dfrac{3}{4}$ $=$ $\dfrac{7}{4}$
   answer

4. $\dfrac{5}{5}$ $+$ $\dfrac{1}{5}$ $=$ ____
   answer

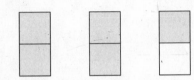

2. $\dfrac{2}{2}$ $+$ $\dfrac{2}{2}$ $+$ $\dfrac{1}{2}$ $=$ ____
   answer

5. $\dfrac{1}{2}$ $+$ $\dfrac{1}{2}$ $+$ $\dfrac{2}{2}$ $=$ ____
   answer

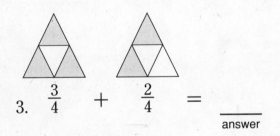

3. $\dfrac{3}{4}$ $+$ $\dfrac{2}{4}$ $=$ ____
   answer

6. $\dfrac{3}{3}$ $+$ $\dfrac{3}{3}$ $+$ $\dfrac{1}{3}$ $=$ ____
   answer

Draw and shade in each of the improper fractions below using the rectangles.

7. $\dfrac{3}{2}$

9. $\dfrac{5}{4}$

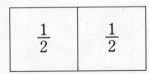

$\dfrac{1}{2}$ $\dfrac{1}{2}$ $\dfrac{1}{2}$ $\dfrac{1}{2}$

shade in

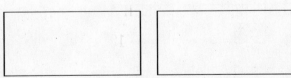

8. $\dfrac{6}{3}$

10. $\dfrac{7}{5}$

5

Mixed Numbers

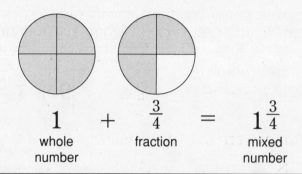

$$1 \quad + \quad \frac{3}{4} \quad = \quad 1\frac{3}{4}$$

whole   fraction   mixed
number   number

> A **mixed number** combines a whole number and a proper fraction.

Write the mixed numbers.

1.

$$1 \quad + \quad 1 \quad + \quad \frac{1}{2} \quad = \quad \underline{2\frac{1}{2}}$$

mixed number

2.

$$1 \quad + \quad \frac{2}{3} \quad = \quad \underline{}$$

mixed number

3.

$$1 \quad + \quad \frac{7}{10} \quad = \quad \underline{}$$

mixed number

4.

$$1 \quad + \quad 1 \quad + \quad \frac{3}{4} \quad = \quad \underline{}$$

mixed number

5. $2 \quad + \quad \frac{5}{7} \quad = \quad \underline{}$

mixed number

6. $5 \quad + \quad \frac{2}{9} \quad = \quad \underline{}$

mixed number

7. $1 \quad + \quad \frac{7}{8} \quad = \quad \underline{}$

mixed number

8. $7 \quad + \quad \frac{1}{2} \quad = \quad \underline{}$

mixed number

9. $3 \quad + \quad \frac{2}{3} \quad = \quad \underline{}$

mixed number

10. $6 \quad + \quad \frac{3}{5} \quad = \quad \underline{}$

mixed number

11. $2 \quad + \quad \frac{1}{6} \quad = \quad \underline{}$

mixed number

12. $1 \quad + \quad \frac{1}{4} \quad = \quad \underline{}$

mixed number

Match the Mixed Numbers

Remember: a mixed number combines a whole number and a proper fraction.

Examples: $2 + \frac{2}{3} = 2\frac{2}{3}$ $9 + \frac{3}{4} = 9\frac{3}{4}$

Match each drawing with the correct mixed number.

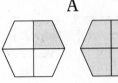
A

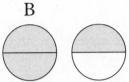
B

1. $\left(2\frac{2}{3}\right)$ 2. $\left(1\frac{3}{4}\right)$

C
_____
letter

_____
letter

C

D

3. $\left(1\frac{1}{4}\right)$ 4. $\left(2\frac{2}{6}\right)$

_____
letter

_____
letter

E

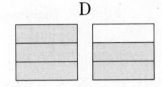
F

5. $\left(1\frac{2}{3}\right)$ 6. $\left(2\frac{1}{4}\right)$

_____
letter

_____
letter

G

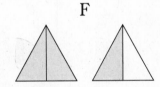
H

7. $\left(2\frac{1}{2}\right)$ 8. $\left(1\frac{1}{2}\right)$

_____
letter

_____
letter

Draw rectangles and shade in each of the mixed numbers below.

9. $1\frac{1}{4}$ — One and one fourth

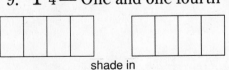
shade in

10. $2\frac{2}{5}$ — Two and two fifths

11. $2\frac{1}{3}$ — Two and one third

12. $2\frac{1}{5}$ — Two and one fifth

7

Writing Fractions and Mixed Numbers

For each of the shaded figures, write the improper fraction and the mixed number.

Improper Fraction   **Mixed Number**

1. $\dfrac{3}{2}$ = 1 $\dfrac{1}{2}$

2. $\dfrac{\Box}{3}$ = $\Box$ $\dfrac{\Box}{\Box}$

3. $\dfrac{\Box}{\Box}$ = $\Box$ $\dfrac{\Box}{\Box}$

4. $\dfrac{\Box}{\Box}$ = $\Box$ $\dfrac{\Box}{\Box}$

5. $\dfrac{\Box}{\Box}$ = $\Box$ $\dfrac{\Box}{\Box}$

6. $\dfrac{\Box}{\Box}$ = $\Box$ $\dfrac{\Box}{\Box}$

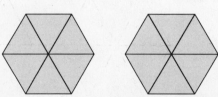

Shade the Mixed Numbers

Shade in the figures to show the mixed numbers.

1. $1\frac{1}{2}$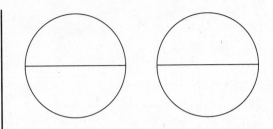

2. $1\frac{3}{4}$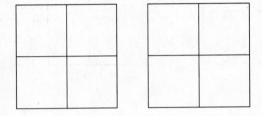

3. $1\frac{3}{5}$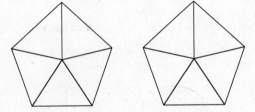

4. $2\frac{3}{8}$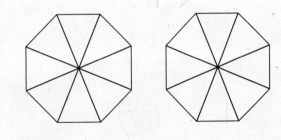

5. $2\frac{5}{6}$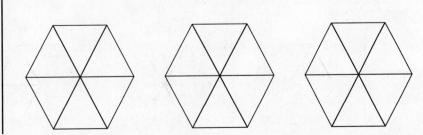

Write the Numbers

Write the fractions, mixed numbers, or whole numbers shown below.

1. If $\frac{1}{6}$ $\frac{1}{6}$ $\frac{1}{6}$ $\frac{1}{6}$ $\frac{1}{6}$ $\frac{1}{6}$ = 1 then 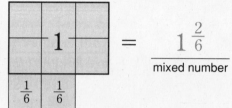 = $1\frac{2}{6}$
    <u>_____</u>
    mixed number

2. If [grid] = 1 then [grid] = _____
    mixed number

3. If [grid] = 1 then [grid] = _____
    fraction

4. If [grid] = 1 then [grid] = _____
    mixed number

5. If $\frac{1}{4}$ $\frac{1}{4}$ $\frac{1}{4}$ $\frac{1}{4}$ = 1 then $\frac{1}{4}$ $\frac{1}{4}$ $\frac{1}{4}$ $\frac{1}{4}$ $\frac{1}{4}$ = $1\frac{1}{4}$
    <u>_____</u>
    mixed number

6. If ☐ ☐ ☐ ☐ = 1 then [squares] = _____
    whole number

7. If ☐ ☐ ☐ ☐ = 1 then ☐ ☐ ☐ = _____
    fraction

8. If ☐ ☐ ☐ ☐ = 1 then [squares] = _____
    mixed number

10

Sets

Match each drawing with the correct answer.

1. $\frac{1}{2}$ of 8

   $\underline{\quad H \quad}$
   letter

2. $\frac{1}{2}$ of 4

   $\underline{\qquad}$
   letter

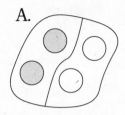 A.

B.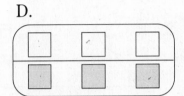

3. $\frac{1}{3}$ of 3

   $\underline{\qquad}$
   letter

4. $\frac{1}{2}$ of 6

   $\underline{\qquad}$
   letter

C.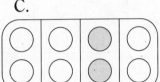

D.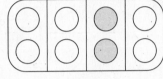

5. $\frac{1}{2}$ of 2

   $\underline{\qquad}$
   letter

6. $\frac{1}{3}$ of 6

   $\underline{\qquad}$
   letter

E.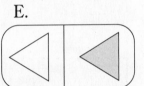

F.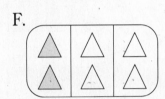

7. $\frac{1}{4}$ of 8

   $\underline{\qquad}$
   letter

8. $\frac{1}{2}$ of 3

   $\underline{\qquad}$
   letter

G.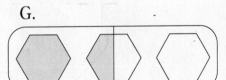

H.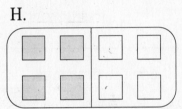

Shade the drawings.

9. $\frac{1}{2}$ of 4

10. $\frac{1}{3}$ of 6

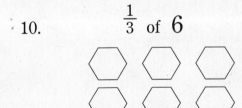

11. $\frac{1}{2}$ of 10

12. $\frac{1}{4}$ of 8

11

More Sets

Match each drawing with the correct answer.

1.   4. $\frac{3}{4}$ of 8

$\underline{\quad E \quad}$
letter

$\underline{\qquad}$
letter

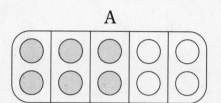
A

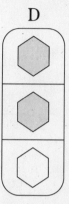
D

2. $\frac{4}{5}$ of 5 5. $\frac{3}{5}$ of 10

$\underline{\qquad}$
letter

$\underline{\qquad}$
letter

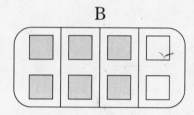
B

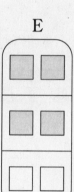
E

3. $\frac{2}{3}$ of 3

$\underline{\qquad}$
letter

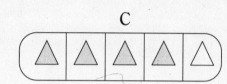
C

Shade the drawings.

6. $\frac{3}{4}$ of 4

8. $\frac{2}{3}$ of 6

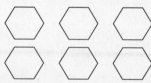

7. $\frac{4}{5}$ of 10

9. $\frac{3}{4}$ of 12

Looking at Sets

1. Shade $\frac{1}{2}$ of the six squares.

2. Circle $\frac{2}{3}$ of the nine Xs.

X X X
X X X
X X X

3. Shade $\frac{1}{4}$ of the twelve squares.

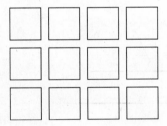

4. Circle $\frac{1}{3}$ of the twelve Xs.

X X X X
X X X X
X X X X

5. Shade $\frac{3}{5}$ of the ten squares.

6. Shade $\frac{4}{5}$ of the ten squares.

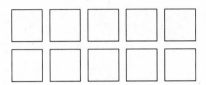

7. Circle $\frac{3}{4}$ of the twelve Xs.

X X X X
X X X X
X X X X

8. Shade $\frac{1}{2}$ of the eight squares.

9. Circle $\frac{5}{6}$ of the six Xs.

X X X X X X

10. Shade $\frac{2}{3}$ of the eighteen squares.

Comparing Fractions with the Same Denominator

1.	2.	3.	4.	5.	6.
Shade $\frac{2}{6}$	Shade $\frac{5}{6}$	Shade $\frac{1}{6}$	Shade $\frac{3}{6}$	Shade $\frac{6}{6}$	Shade $\frac{4}{6}$

Arrange the fractions above from smallest to largest.

7. _____ _____ _____ _____ _____ _____

    smallest   largest

8. Shade $\frac{1}{4}$	9. Shade $\frac{3}{4}$	10. Shade $\frac{2}{4}$	11. Shade $\frac{0}{4}$	12. Shade $\frac{4}{4}$

Arrange the fractions above from smallest to largest.

13. _____ _____ _____ _____ _____

     smallest   largest

If two fractions have the same denominator:

14.   The fraction with the larger numerator is _____.

                                                 smaller or larger

15.   The fraction with the smaller numerator is _____.

                                               smaller or larger

Comparing Fractions with the Same Numerator

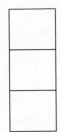

1. Shade
$\frac{2}{3}$

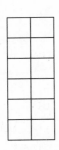

2. Shade
$\frac{2}{12}$

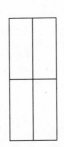

3. Shade
$\frac{2}{4}$

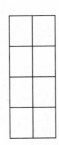

4. Shade
$\frac{2}{8}$

Arrange the fractions above from smallest to largest.

5. _____   _____   _____   _____
   smallest   largest

_____

6. Shade
$\frac{3}{4}$

7. Shade
$\frac{3}{8}$

8. Shade
$\frac{3}{5}$

9. Shade
$\frac{3}{3}$

Arrange the fractions above from smallest to largest.

10. _____   _____   _____   _____
    smallest   largest

_____

If two fractions have the same numerator:

11.   The fraction with the larger denominator is _____.
                                              smaller or larger

12.   The fraction with the smaller denominator is _____.
                                              smaller or larger

15

Shade and Compare

1. $\frac{2}{5}$

2. $\frac{1}{2}$

3. Larger fraction _____

4. Smaller fraction _____

5. Shade $\frac{5}{6}$

6. Shade $\frac{1}{2}$

7. Larger fraction _____

8. Smaller fraction _____

9. Shade $\frac{4}{5}$

10. Shade $\frac{1}{2}$

11. Larger fraction _____

12. Smaller fraction _____

13. Shade $\frac{1}{3}$

14. Shade $\frac{1}{2}$

15. Larger fraction _____

16. Smaller fraction _____

17. Shade $\frac{5}{6}$

18. Shade $\frac{4}{6}$

19. Larger fraction _____

20. Smaller fraction _____

21. Shade $\frac{3}{4}$

22. Shade $\frac{6}{8}$

23. What do you notice about these two fractions? _____

24. Shade $\frac{5}{3}$

25. Shade $\frac{3}{2}$

26. Larger fraction _____

27. Smaller fraction _____

28. Shade $\frac{2}{3}$

29. Shade $\frac{5}{6}$

30. Larger fraction _____

31. Smaller fraction _____

Order the Fractions

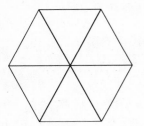

1. Shade $\frac{5}{6}$

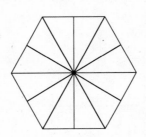

2. Shade $\frac{11}{12}$

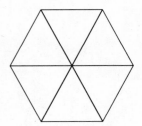

3. Shade $\frac{1}{6}$

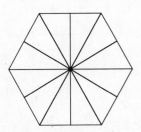

4. Shade $\frac{1}{12}$

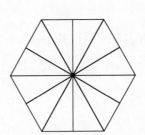

5. Shade $\frac{5}{12}$

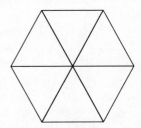

6. Shade $\frac{3}{6}$

Compare the size of the shaded parts in each drawing. Arrange the fractions above in order from smallest to largest.

7. _____ _____ _____ _____ _____ _____

   smallest   largest

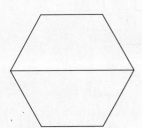

8. Shade $\frac{1}{2}$

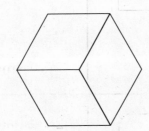

9. Shade $\frac{2}{3}$

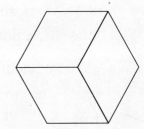

10. Shade $\frac{1}{3}$

Compare the size of the shaded parts in each drawing. Arrange the fractions above in order from smallest to largest.

11. _____ _____ _____

   smallest   largest

What Are Equivalent Fractions?

A   B

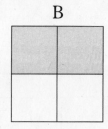

1. Are the two squares A and B the same size? _____

2. What fraction is shaded in square A? _____

3. What fraction is shaded in square B? _____

4. Are the unshaded parts in both A and B the same size? _____

5. Are the shaded parts in A and B the same size? _____

6. Is $\frac{1}{2} = \frac{2}{4}$? _____

Fractions that have the same (equal) value are **equivalent fractions**.

Look at the shaded parts in the figures below and fill in the equivalent fractions.

7.
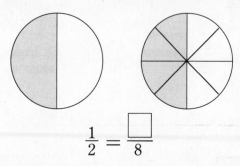

$$\frac{1}{2} = \frac{\square}{8}$$

9.
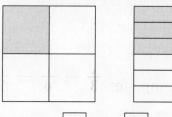

$$\frac{\square}{\square} = \frac{\square}{\square}$$

8.
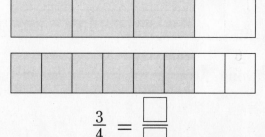

$$\frac{3}{4} = \frac{\square}{\square}$$

10.
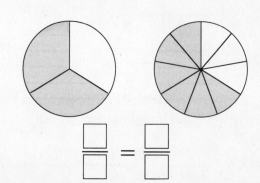

$$\frac{\square}{\square} = \frac{\square}{\square}$$

Shading Equivalent Fractions

1. Shade $\frac{1}{2}$

2. Shade $\frac{2}{4}$

3. Shade $\frac{4}{8}$

4. Are the unshaded parts the same size in each of the rectangles? _____

5. Are the shaded parts in each rectangle the same size? _____

6. Complete: $\frac{1}{2} = \frac{}{4} = \frac{}{8}$

7.

Shade $\frac{1}{3}$

8.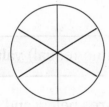

Shade $\frac{2}{6}$

9.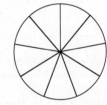

Shade $\frac{3}{9}$

10. Complete: $\frac{1}{3} = \frac{}{6} = \frac{}{9}$

11.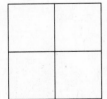

Shade $\frac{3}{4}$

12.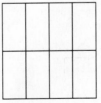

Shade $\frac{6}{8}$

13.

Shade $\frac{9}{12}$

14. Complete: $\frac{3}{4} = \frac{}{8} = \frac{}{12}$

19

Writing Equivalent Fractions

1							
$\frac{1}{2}$				$\frac{1}{2}$			
$\frac{1}{4}$		$\frac{1}{4}$		$\frac{1}{4}$		$\frac{1}{4}$	
$\frac{1}{8}$	$\frac{1}{8}$	$\frac{1}{8}$	$\frac{1}{8}$	$\frac{1}{8}$	$\frac{1}{8}$	$\frac{1}{8}$	$\frac{1}{8}$
$\frac{1}{16}$ $\frac{1}{16}$	$\frac{1}{16}$ $\frac{1}{16}$	$\frac{1}{16}$ $\frac{1}{16}$	$\frac{1}{16}$ $\frac{1}{16}$	$\frac{1}{16}$ $\frac{1}{16}$	$\frac{1}{16}$ $\frac{1}{16}$	$\frac{1}{16}$ $\frac{1}{16}$	$\frac{1}{16}$ $\frac{1}{16}$

EXAMPLE 1

$\frac{1}{2}$	
$\frac{1}{4}$	$\frac{1}{4}$

$$\frac{1}{2} = \frac{2}{4}$$

EXAMPLE 2

$\frac{1}{4}$	
$\frac{1}{8}$	$\frac{1}{8}$
$\frac{1}{16}$ $\frac{1}{16}$	$\frac{1}{16}$ $\frac{1}{16}$

$$\frac{1}{4} = \frac{2}{8} = \frac{4}{16}$$

Use the chart to fill in the equivalent fractions below.

1. $\frac{1}{2} = \frac{}{4}$

2. $\frac{1}{4} = \frac{}{8}$

3. $\frac{1}{8} = \frac{}{16}$

4. $\frac{1}{2} = \frac{}{8}$

5. $\frac{3}{4} = \frac{}{8}$

6. $\frac{4}{8} = \frac{}{4}$

7. $\frac{2}{16} = \frac{}{8}$

8. $\frac{3}{8} = \frac{}{16}$

9. $1 = \frac{}{8}$

10. $\frac{6}{16} = \frac{}{8}$

11. $\frac{5}{8} = \frac{}{16}$

12. $\frac{4}{16} = \frac{}{4}$

13. $\frac{10}{16} = \frac{}{8}$

14. $1 = \frac{}{2}$

15. $\frac{3}{4} = \frac{}{16}$

16. $\frac{8}{16} = \frac{}{2}$

17. $\frac{3}{4} = \frac{}{8} = \frac{}{16}$

18. $\frac{1}{2} = \frac{}{4} = \frac{}{16}$

19. $1 = \frac{}{2} = \frac{}{8}$

20. $\frac{4}{16} = \frac{}{8} = \frac{}{4}$

21. $\frac{16}{16} = \frac{}{8} = \frac{}{4}$

22. $\frac{4}{4} = \frac{}{2} = \frac{}{1}$

23. $\frac{2}{8} = \frac{}{4} = \frac{}{16}$

24. $\frac{12}{16} = \frac{}{4} = \frac{}{8}$

20

More Equivalent Fractions

You cannot use charts all the time to compare fractions. To find equivalent fractions, multiply both the numerator and denominator by the same number.

A fraction with the same numerator and denominator is equal to 1.

$$\frac{2}{5} = \frac{2}{5} \times \boxed{1\frac{2}{2}} = \frac{2 \times 2}{5 \times 2} = \frac{4}{10}$$

Different form, same value.

1. $\dfrac{1}{3} = \dfrac{1 \times \boxed{5}}{3 \times \boxed{5}} = \dfrac{\boxed{}}{15}$

7. $\dfrac{3}{7} = \dfrac{3 \times 5}{7 \times 5} = \dfrac{\boxed{}}{35}$

2. $\dfrac{3}{4} = \dfrac{3 \times \boxed{3}}{4 \times \boxed{3}} = \dfrac{\boxed{}}{12}$

8. $\dfrac{5}{12} = \dfrac{5 \times 2}{12 \times 2} = \dfrac{\boxed{}}{24}$

3. $\dfrac{5}{8} = \dfrac{5 \times \boxed{4}}{8 \times \boxed{4}} = \dfrac{\boxed{}}{32}$

9. $\dfrac{5}{6} = \dfrac{5 \times 3}{6 \times 3} = \dfrac{\boxed{}}{18}$

4. $\dfrac{2}{3} = \dfrac{2 \times 4}{3 \times 4} = \dfrac{\boxed{}}{12}$

10. $\dfrac{7}{9} = \dfrac{7 \times 7}{9 \times 7} = \dfrac{\boxed{}}{63}$

5. $\dfrac{1}{2} = \dfrac{1 \times 7}{2 \times 7} = \dfrac{\boxed{}}{14}$

11. $\dfrac{3}{8} = \dfrac{3 \times 7}{8 \times 7} = \dfrac{\boxed{}}{56}$

6. $\dfrac{3}{5} = \dfrac{3 \times 4}{5 \times 4} = \dfrac{\boxed{}}{20}$

12. $\dfrac{7}{12} = \dfrac{7 \times 4}{12 \times 4} = \dfrac{\boxed{}}{48}$

Using Fractions Equal to 1

Multiplying a number or fraction by 1 does not change its value.

To find an equivalent fraction:

$$\frac{1}{3} = \frac{1 \times 4}{3 \times 4} = \frac{4}{12}$$

Think: what number can you multiply 1 by to equal 4?

Think: what number can you multiply 3 by to equal 12?

so $\frac{1}{3}$ is equal to $\frac{4}{12}$

Fill in a fraction (with a value of 1) that will complete each operation.

1. $\frac{2}{5} = \dfrac{2 \times \boxed{3}}{5 \times \square} = \frac{6}{15}$ so $\dfrac{\boxed{2}}{\boxed{5}}$ is equal to $\dfrac{\square}{\square}$

2. $\frac{3}{4} = \dfrac{3 \times \square}{4 \times \boxed{4}} = \frac{12}{16}$ so $\dfrac{\square}{\square}$ is equal to $\dfrac{\boxed{12}}{\boxed{16}}$

3. $\frac{5}{6} = \dfrac{5 \times \square}{6 \times \square} = \frac{10}{12}$ so $\dfrac{\square}{\square}$ is equal to $\dfrac{\square}{\square}$

4. $\frac{2}{7} = \dfrac{2 \times \square}{7 \times \square} = \frac{4}{14}$ so $\dfrac{\square}{\square}$ is equal to $\dfrac{\square}{\square}$

5. $\frac{1}{8} = \dfrac{1 \times \square}{8 \times \square} = \frac{3}{24}$ so $\dfrac{\square}{\square}$ is equal to $\dfrac{\square}{\square}$

6. $\frac{2}{9} = \dfrac{2 \times \square}{9 \times \square} = \frac{6}{27}$ so $\dfrac{\square}{\square}$ is equal to $\dfrac{\square}{\square}$

Find the Numerators

Remember to multiply both the numerator and the denominator by the same number.

$$\frac{3}{4} = \frac{9}{12} \quad \text{because} \quad \frac{3}{4} = \frac{3 \times}{4 \times} \boxed{\frac{3}{3}} = \frac{9}{12}$$

Fill in the correct numerators to make the fractions equal in value.

1. $\frac{1}{4} = \frac{}{20}$ because $\frac{1 \times \Box}{4 \times \Box} = \frac{}{20}$

2. $\frac{3}{5} = \frac{}{15}$ because $\frac{3 \times \Box}{5 \times \Box} = \frac{}{15}$

3. $\frac{4}{7} = \frac{}{21}$ because $\frac{4 \times \Box}{7 \times \Box} = \frac{}{21}$

4. $\frac{1}{3} = \frac{}{15}$ because $\frac{1 \times \Box}{3 \times \Box} = \frac{}{15}$

5. $\frac{5}{6} = \frac{}{12}$ because $\frac{5 \times \Box}{6 \times \Box} = \frac{}{12}$

6. $\frac{2}{9} = \frac{}{36}$ because $\frac{2 \times \Box}{9 \times \Box} = \frac{}{36}$

7. $\frac{2}{3} = \frac{}{18}$ because $\frac{2 \times \Box}{3 \times \Box} = \frac{}{18}$

8. $\frac{3}{4} = \frac{}{48}$ because $\frac{3 \times \Box}{4 \times \Box} = \frac{}{48}$

Practice

To find equivalent fractions, multiply the numerator and the denominator of the fraction by the same number.

Complete the fractions.

1. $\dfrac{1 \times 4}{3 \times 4} = \dfrac{4}{12}$

2. $\dfrac{3 \times 3}{4 \times 3} = \dfrac{9}{}$

3. $\dfrac{1}{2} = \dfrac{2}{}$

4. $\dfrac{2}{3} = \dfrac{}{9}$

5. $\dfrac{1}{4} = \dfrac{4}{}$

6. $\dfrac{1}{5} = \dfrac{}{45}$

7. $\dfrac{1}{3} = \dfrac{}{27}$

8. $\dfrac{1}{7} = \dfrac{2}{}$

9. $\dfrac{2}{5} = \dfrac{4}{}$

10. $\dfrac{1}{6} = \dfrac{}{42}$

11. $\dfrac{7}{10} = \dfrac{21}{}$

12. $\dfrac{5}{7} = \dfrac{35}{}$

13. $\dfrac{7}{12} = \dfrac{}{24}$

14. $\dfrac{11}{50} = \dfrac{}{100}$

15. $\dfrac{9}{20} = \dfrac{18}{}$

16. $\dfrac{7}{15} = \dfrac{14}{}$

17. $\dfrac{11}{14} = \dfrac{}{28}$

18. $\dfrac{7}{8} = \dfrac{}{32}$

19. $\dfrac{5}{9} = \dfrac{25}{}$

20. $\dfrac{1}{4} = \dfrac{}{12}$

Less Than or Greater Than

Find the missing numerators and then compare the fractions using the symbols
$<$ (less than) or $>$ (greater than).

1. a) $\dfrac{1}{4} = \dfrac{2}{8}$

    b) $\dfrac{1}{2} = \dfrac{4}{8}$

    c) $\dfrac{1}{4} \underline{\ <\ } \dfrac{1}{2}$
       $\left(\dfrac{2}{8}\right)$ $\left(\dfrac{4}{8}\right)$

2. a) $\dfrac{1}{3} = \dfrac{}{24}$

    b) $\dfrac{3}{8} = \dfrac{}{24}$

    c) $\dfrac{1}{3} \underline{\quad} \dfrac{3}{8}$

3. a) $\dfrac{5}{6} = \dfrac{}{24}$

    b) $\dfrac{3}{4} = \dfrac{}{24}$

    c) $\dfrac{5}{6} \underline{\quad} \dfrac{3}{4}$

4. a) $\dfrac{2}{3} = \dfrac{}{15}$

    b) $\dfrac{4}{5} = \dfrac{}{15}$

    c) $\dfrac{2}{3} \underline{\quad} \dfrac{4}{5}$

5. a) $\dfrac{1}{2} = \dfrac{}{6}$

    b) $\dfrac{2}{3} = \dfrac{}{6}$

    c) $\dfrac{1}{2} \underline{\quad} \dfrac{2}{3}$

6. a) $\dfrac{3}{4} = \dfrac{}{12}$

    b) $\dfrac{4}{6} = \dfrac{}{12}$

    c) $\dfrac{3}{4} \underline{\quad} \dfrac{4}{6}$

Compare Using the Number Line

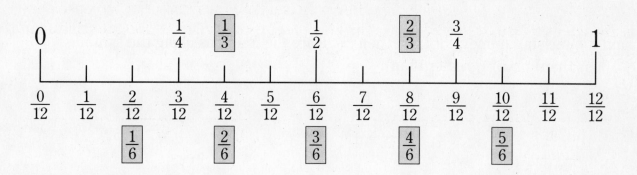

Find both fractions on the number line and compare using the symbols < (less than), > (greater than), or = (equal to). Remember: the closer a fraction is to zero, the smaller its value.

1. $\dfrac{1}{4}$ $\underline{<}$ $\dfrac{1}{3}$

2. $\dfrac{4}{6}$ ___ $\dfrac{2}{3}$

3. $\dfrac{5}{6}$ ___ $\dfrac{7}{12}$

4. $\dfrac{11}{12}$ ___ $\dfrac{3}{4}$

5. $\dfrac{5}{12}$ ___ $\dfrac{7}{12}$

6. $\dfrac{6}{12}$ ___ $\dfrac{3}{6}$

7. $\dfrac{5}{6}$ ___ $\dfrac{5}{12}$

8. $\dfrac{1}{2}$ ___ $\dfrac{7}{12}$

9. $\dfrac{1}{6}$ ___ $\dfrac{2}{12}$

10. $\dfrac{3}{4}$ ___ $\dfrac{8}{12}$

11. $\dfrac{2}{6}$ ___ $\dfrac{1}{12}$

12. $\dfrac{9}{12}$ ___ $\dfrac{5}{6}$

Connect Tags to Lines

Connect each tag to the number line. Begin by counting the number of parts each line is divided into. You may need to use the number line on page 26 to find the equivalent fractions. Write the correct letter below each tag.

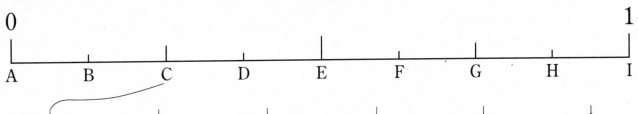

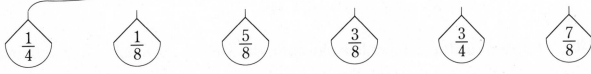

1. $\underset{\text{letter}}{\text{C}}$ $\underset{\text{letter}}{\quad}$ $\underset{\text{letter}}{\quad}$ $\underset{\text{letter}}{\quad}$ $\underset{\text{letter}}{\quad}$ $\underset{\text{letter}}{\quad}$

$\left(\frac{1}{4} = \frac{2}{8}\right)$ $\left(\frac{3}{4} = \frac{?}{8}\right)$

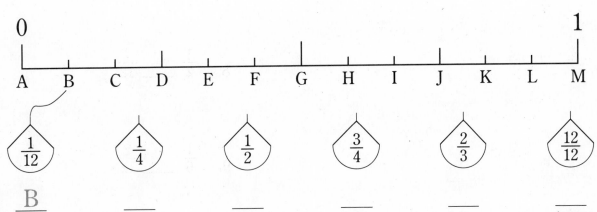

2. $\underset{\text{letter}}{\text{B}}$ $\underset{\text{letter}}{\quad}$ $\underset{\text{letter}}{\quad}$ $\underset{\text{letter}}{\quad}$ $\underset{\text{letter}}{\quad}$ $\underset{\text{letter}}{\quad}$

$\left(\frac{1}{4} = \frac{?}{12}\right)$

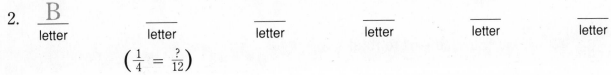

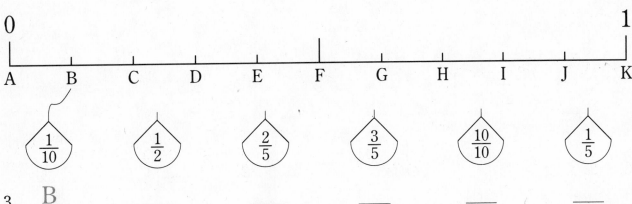

3. $\underset{\text{letter}}{\text{B}}$ $\underset{\text{letter}}{\quad}$ $\underset{\text{letter}}{\quad}$ $\underset{\text{letter}}{\quad}$ $\underset{\text{letter}}{\quad}$ $\underset{\text{letter}}{\quad}$

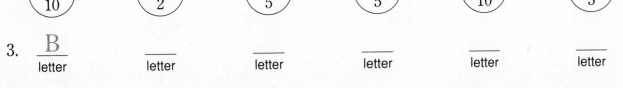

Fractions Show Comparisons

A fraction can be thought of as a comparison. Write the fraction for each drawing below.

Fraction

1. $\frac{3}{5}$ Shaded squares to total squares

2. $\underline{}$ Indicated units to total units

3. $\underline{}$ Shaded circle to total circles

4. $\underline{}$ Shaded part to total

5. $\underline{}$ Shaded part to total

6. $\underline{}$ Shaded part to total

7. $\underline{}$ Shaded boxes to total boxes

8. $\underline{}$ Shaded part to total

More Comparisons

A fraction can be thought of as a comparison. For example, a basketball team can compare its number of wins to the total number of games played. If a team plays 10 games and wins 7, the fraction of wins to total games would be:

$$\frac{\text{part}}{\text{total}} = \frac{7}{10} \quad \text{of the games were won}$$

Write these comparisons as fractions.

1. 30 students
    15 earned an "A" in math

    $$\frac{\text{part}}{\text{total}} = \frac{\boxed{}}{30} \quad \text{received an "A"}$$

2. 10 hits
    20 times at bat

    $$\frac{\text{part}}{\text{total}} = \frac{\boxed{}}{\boxed{}} \quad \text{were hits}$$

3. A rifleman fired 10 shots.
    He hit the target 7 times.

    $$\frac{\text{part}}{\text{total}} = \frac{\boxed{}}{\boxed{}} \quad \text{hit the target}$$

4. Joan's book has 125 pages.
    She read 25 pages.

    $$\frac{\text{part}}{\text{total}} = \frac{\boxed{}}{\boxed{}} \quad \text{of the book was read}$$

5. 6 coins
    2 are pennies

    $$\frac{\text{part}}{\text{total}} = \frac{\boxed{}}{\boxed{}} \quad \text{of the coins are pennies}$$

6. A store sold 10 radios.
    2 were returned.

    $$\frac{\text{part}}{\text{total}} = \frac{\boxed{}}{\boxed{}} \quad \text{of the radios were returned}$$

7. The cake was divided into 12 equal pieces. We ate 2 of them.

    $$\frac{\text{part}}{\text{total}} = \frac{\boxed{}}{\boxed{}} \quad \text{of the cake was eaten}$$

8. It rained 3 out of 4 days.

    $$\frac{\text{part}}{\text{total}} = \frac{\boxed{}}{\boxed{}} \quad \text{of the days were rainy}$$

Fraction Review

1. Which is a correct drawing for $\frac{1}{4}$?

$\overline{\hspace{1cm}}$
letter

  A   B   C

2. Shade $\frac{4}{5}$

3. Which is larger $\frac{3}{5}$ or $\frac{3}{8}$?

$\overline{\hspace{1cm}}$
answer

4. Shade $\frac{3}{4}$ of 8

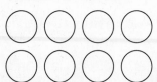

5. Circle $\frac{1}{3}$ of 9

6. Shade $\frac{5}{4}$ which equals $\overline{\hspace{2cm}}$
mixed number

7. $\frac{3}{4} = \frac{}{20}$

8. $\frac{5}{8} = \frac{10}{}$

9. 4 times at bat
   1 hit

   ☐ What fraction
   ☐ were hits?

10. What fraction of the line is shown at point A?

$\overline{\hspace{1cm}}$
answer

0   A   1

Divisibility Rule for 2

Sometimes it is helpful to know if a number is divisible by 2.

$$\begin{array}{r} 3 \\ 2\overline{)6} \\ \underline{6} \\ 0 \end{array}$$

6 **is** divisible by 2 because it divides exactly with a remainder of 0.

$$\begin{array}{r} 2 \\ 2\overline{)5} \\ \underline{4} \\ 1 \end{array}$$

5 **is not** divisible by 2 because it does not divide exactly. The remainder is not 0.

Any number that ends in 0, 2, 4, 6 or 8 is divisible by 2.

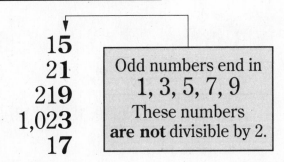

68
20
194
1,002
66

Even numbers end in
0, 2, 4, 6, 8
These numbers
are divisible by 2.

15
21
219
1,023
17

Odd numbers end in
1, 3, 5, 7, 9
These numbers
are not divisible by 2.

Circle the numbers that are divisible by 2.

1.	22	7.	107	13.	655
2.	71	8.	39	14.	938
3.	15	9.	75	15.	1,004
4.	293	10.	212	16.	86
5.	96	11.	200	17.	95
6.	104	12.	93	18.	100

Divisibility Rule for 3

Sometimes it is helpful to know if a number is divisible by 3.

$$3\overline{)12} \quad \frac{4}{}$$

12 **is** divisible by 3 because it divides exactly with a remainder of 0.

$$3\overline{)11} \quad \frac{3}{}$$

11 **is not** divisible by 3 because it does not divide exactly. The remainder is not 0.

If the sum of the digits in any number is divisible by 3, the number is divisible by 3.

$$3\overline{)\boxed{54}}$$
$5 + 4 = 9$

$$3\overline{)\boxed{642}}$$
$6 + 4 + 2 = 12$

$$3\overline{)\boxed{567}}$$
$5 + 6 + 7 = 18$

The sums of the digits (9, 12, and 18) are all divisible by 3, so 54, 642, and 567 are all divisible by 3.

Find the sum of the digits and then circle the numbers that are divisible by 3.

Think: 2 + 7 =

1.	(27)	9	7.	46	☐	13.	66	☐
2.	35	☐	8.	57	☐	14.	54	☐
3.	60	☐	9.	21	☐	15.	621	☐
4.	56	☐	10.	24	☐	16.	304	☐
5.	92	☐	11.	19	☐	17.	555	☐
6.	31	☐	12.	85	☐	18.	206	☐

Divisibility Rule for 5

Sometimes it is helpful to know if a number is divisible by 5.

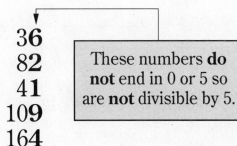

$$\begin{array}{r} 6 \\ 5\overline{)\,3\,0} \\ 3\,0 \\ \hline 0 \end{array}$$

30 **is** divisible by 5 because it divides exactly with a remainder of 0.

$$\begin{array}{r} 5 \\ 5\overline{)\,2\,9} \\ 2\,5 \\ \hline 4 \end{array}$$

29 **is not** divisible by 5 because it does not divide exactly. The remainder is not 0.

> Any number that ends in 0 or 5 is divisible by 5.

30
85
290
1,000
955

All end in 0 or 5 and **are** divisible by 5.

36
82
41
109
164

These numbers **do not** end in 0 or 5 so are **not** divisible by 5.

Circle all the numbers that are divisible by 5.

1.	48	7.	62	13.	180
2.	(55)	8.	35	14.	105
3.	62	9.	80	15.	69
4.	93	10.	572	16.	908
5.	15	11.	24	17.	370
6.	50	12.	175	18.	90

Divisibility Rule for 10

Sometimes it is helpful to know if a number is divisible by 10.

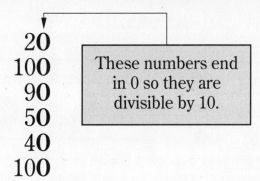

30 **is** divisible by 10 because it divides exactly with a remainder of 0.

38 **is not** divisible by 10 because it does not divide exactly. The remainder is not 0.

Any number that ends in 0 is divisible by 10.

20
100
90
50
40
100

These numbers end in 0 so they are divisible by 10.

21
106
1,005
98
33
45

These numbers **do not** end in 0 so they are **not** divisible by 10.

Circle all the numbers that are divisible by 10.

1.	30	7.	638	13.	556
2.	38	8.	70	14.	200
3.	92	9.	815	15.	681
4.	57	10.	309	16.	93
5.	65	11.	88	17.	755
6.	105	12.	905	18.	450

Divisibility Practice

Check the column if the number is divisible by 2, 3, 5, or 10.

	Number	Divisible by 2	3	5	10
1.	40	✔		✔	✔
2.	144				
3.	94				
4.	540				
5.	1,000				
6.	29				
7.	45				
8.	85				
9.	342				
10.	70				
11.	100				
12.	65				
13.	38				
14.	153				
15.	384				
16.	89				
17.	420				
18.	5,546				

Finding Factors

Use factors to simplify fractions.

Factors are:
- all the numbers that can be multiplied to find a given number
- the number itself and 1

To find all the factors of 15, **think:**

$$\underline{\quad 1 \quad} \times \underline{\quad 15 \quad} = \underline{\quad 15 \quad}$$
factor $\qquad$ factor $\qquad$ product

$$\underline{\quad 3 \quad} \times \underline{\quad 5 \quad} = \underline{\quad 15 \quad}$$
factor $\qquad$ factor $\qquad$ product

1 × 15

3 × 5

$$\underline{\quad 1 \quad} \qquad \underline{\quad 3 \quad} \qquad \underline{\quad 5 \quad} \qquad \underline{\quad 15 \quad}$$
factor $\qquad$ factor $\qquad$ factor $\qquad$ factor

1. List all the factors for 15. $\underline{\ 1\ }$, _____ , _____ , $\underline{\ 15\ }$

2. How many factors does 15 have? _____

To find all the factors of 20, **think:**

$$1 \times 20 = 20$$
$$2 \times 10 = 20$$
$$4 \times 5 = 20$$

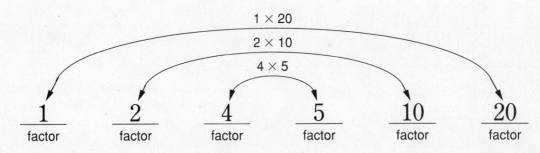

3. List all the factors for 20. $\underline{\ 1\ }$, _____ , _____ , _____ , _____ , $\underline{\ 20\ }$

4. How many factors does 20 have? _____

Name the Factors

Find all the factors of 18.

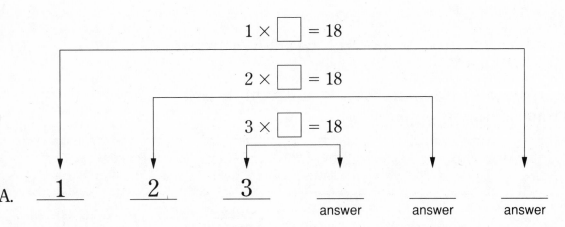

$$1 \times \boxed{} = 18$$

$$2 \times \boxed{} = 18$$

$$3 \times \boxed{} = 18$$

A. ___1___ , ___2___ , ___3___ , _____ , _____ , _____
                                     answer   answer   answer

B. How many factors does 18 have? _____
                                      answer

Write all of the factors for each number.

1. Name all the factors for 12: ____ , __2__ , ____ , ____ , ____ , __12__

list only once

2. Name all the factors for 9: ____ , __3__ , ____

3. Name all the factors for 10: ____ , ____ , ____ , ____

4. Name all the factors for 28: ____ , ____ , ____ , ____ , ____ , ____

5. Name all the factors for 8: ____ , ____ , ____ , ____

6. Name all the factors for 4: ____ , ____ , ____

7. Name all the factors for 7: ____ , ____

8. Name all the factors for 16: ____ , ____ , ____ , ____ , ____

9. Name all the factors for 24: ____ , ____ , ____ , ____ , ____ , ____ , ____ , ____

10. Name all the factors for 30: ____ , ____ , ____ , ____ , ____ , ____ , ____ , ____

Greatest Common Factor

Find the greatest common factor (GCF) for 8 and 12.

The factors of 8 are: $\boxed{1}$ $\boxed{2}$ $\boxed{4}$ 8

The factors of 12 are: $\boxed{1}$ $\boxed{2}$ 3 $\boxed{4}$ 6 12

The factors 1, 2, and 4 are factors of both 8 and 12 and are called common factors. The number 4 is the greatest common factor of 8 and 12.

> The largest factor that two or more numbers have in common is the **greatest common factor.**

Find the greatest common factor (GCF).

1. 4: ____ , ____ , ____
   factors of 4

2. 8: ____ , ____ , ____ , ____
   factors of 8

3. Common factors of 4 and 8: ____ , ____ , ____

4. Greatest common factor of 4 and 8: ____

5. 6: ____ , ____ , ____ , ____
   factors of 6

6. 9: ____ , ____ , ____
   factors of 9

7. Common factors of 6 and 9: ____ , ____

8. Greatest common factor of 6 and 9: ____

9. 16: ____ , ____ , ____ , ____ , ____
   factors of 16

10. 20: ____ , ____ , ____ , ____ , ____ , ____
    factors of 20

11. Common factors of 16 and 20: ____ , ____ , ____

12. Greatest common factor of 16 and 20: ____

Find the Greatest Common Factor

Find the greatest common factor (GCF).

1. 10: ___ , ___ , ___ , ___
   factors of 10

2. 15: ___ , ___ , ___ , ___
   factors of 15

3. Common factors of 10 and 15: ___ , ___

4. Greatest common factor of 10 and 15: ___

5. 8: ___ , ___ , ___ , ___
   factors of 8

6. 12: ___ , ___ , ___ , ___ , ___ , ___
   factors of 12

7. Common factors of 8 and 12: ___ , ___ , ___

8. Greatest common factor of 8 and 12: ___

9. 14: ___ , ___ , ___ , ___
   factors of 14

10. 24: ___ , ___ , ___ , ___ , ___ , ___ , ___ , ___
    factors of 24

11. Common factors of 14 and 24: ___ , ___

12. Greatest common factor of 14 and 24: ___

13. 21: ___ , ___ , ___ , ___
    factors of 21

14. 36: ___ , ___ , ___ , ___ , ___ , ___ , ___ , ___ , ___
    factors of 36

15. Common factors of 21 and 36: ___ , ___

16. Greatest common factor of 21 and 36: ___

Factors with Fractions

You may need to find common factors between the numerator and the denominator of a fraction.

You don't always need to make a list of factors. Think of all whole numbers that divide exactly into both the numerator and the denominator. These numbers will be the common factors.

$\frac{6}{12}$

A. The common factors of 6 and 12 are: __1__ , ____ , ____ , ____

B. The greatest common factor (GCF) is: ____

Find the greatest common factor for the numerator and denominator in each fraction.

$\frac{6}{15}$

1. The common factors of 6 and 15 are: ____ , ____

2. The greatest common factor (GCF) is: ____

$\frac{10}{30}$

3. The common factors of 10 and 30 are: ____ , ____ , ____ , ____

4. The greatest common factor (GCF) is: ____

$\frac{10}{16}$

5. The common factors of 10 and 16 are: ____ , ____

6. The greatest common factor (GCF) is: ____

$\frac{8}{16}$

7. The common factors of 8 and 16 are: ____ , ____ , ____ , ____

8. The greatest common factor (GCF) is: ____

$\frac{7}{21}$

9. The common factors of 7 and 21 are: ____ , ____

10. The greatest common factor (GCF) is: ____

Apply Your Skills

Find the greatest common factor (GCF) for each fraction.

1. $\frac{6}{12}$ GCF = ____

2. $\frac{3}{15}$ GCF = ____

3. $\frac{4}{6}$ GCF = ____

4. $\frac{5}{30}$ GCF = ____

5. $\frac{2}{14}$ GCF = ____

6. $\frac{14}{21}$ GCF = ____

7. $\frac{8}{24}$ GCF = ____

8. $\frac{9}{36}$ GCF = ____

9. $\frac{8}{10}$ GCF = ____

10. $\frac{6}{15}$ GCF = ____

11. $\frac{2}{4}$ GCF = ____

12. $\frac{6}{24}$ GCF = ____

13. $\frac{3}{36}$ GCF = ____

14. $\frac{12}{15}$ GCF = ____

15. $\frac{9}{12}$ GCF = ____

16. $\frac{15}{18}$ GCF = ____

17. $\frac{20}{30}$ GCF = ____

18. $\frac{25}{40}$ GCF = ____

19. $\frac{5}{30}$ GCF = ____

20. $\frac{9}{15}$ GCF = ____

21. $\frac{9}{18}$ GCF = ____

22. $\frac{12}{36}$ GCF = ____

23. $\frac{11}{22}$ GCF = ____

24. $\frac{14}{35}$ GCF = ____

Simplify Fractions

Sometimes you need to simplify fractions or reduce them to their lowest terms.

> To simplify a fraction, divide both the numerator and the denominator by the greatest common factor.

$$\frac{6}{9} \div \frac{3}{3} = \frac{2}{3}$$ Think: GCF of 6 and 9 is 3.

$$\frac{4}{20} \div \frac{4}{4} = \frac{1}{5}$$ Think: GCF of 4 and 20 is 4.

Find the greatest common factor and simplify.

1. $\frac{14}{49}$ GCF = __7__

   $\frac{14}{49} \div \frac{7}{7} =$ _____
   finish

2. $\frac{4}{12}$ GCF = _____

   $\frac{4}{12} \div \frac{\quad}{\quad} =$ _____
   simplified fraction

3. $\frac{24}{32}$ GCF = _____

   _____
   simplified fraction

4. $\frac{20}{30}$ GCF = _____

   _____
   simplified fraction

5. $\frac{11}{33}$ GCF = _____

   _____
   simplified fraction

6. $\frac{18}{45}$ GCF = _____

   _____
   simplified fraction

7. $\frac{25}{30}$ GCF = _____

   _____
   simplified fraction

8. $\frac{30}{45}$ GCF = _____

   _____
   simplified fraction

9. $\frac{32}{36}$ GCF = _____

   _____
   simplified fraction

10. $\frac{26}{39}$ GCF = _____

    _____
    simplified fraction

Simplify

Find the greatest common factor and simplify each fraction.

1. $\dfrac{6}{12} \div \dfrac{6}{6} = \dfrac{1}{2}$

    GCF = 6

2. $\dfrac{3}{15} \div \boxed{1} = \dfrac{1}{5}$

3. $\dfrac{4}{6} \div \dfrac{2}{2} = \underline{}$

4. $\dfrac{5}{30} \div \boxed{} = \underline{}$

5. $\dfrac{2}{14} \div \boxed{} = \underline{}$

6. $\dfrac{14}{21} \div \boxed{} = \underline{}$

7. $\dfrac{8}{24} \div \boxed{} = \underline{}$

8. $\dfrac{9}{36} \div \boxed{} = \underline{}$

9. $\dfrac{8}{10} \div \dfrac{2}{2} = \underline{}$

10. $\dfrac{6}{15} \div \boxed{1} = \underline{}$

11. $\dfrac{2}{4} \div \boxed{} = \underline{}$

12. $\dfrac{6}{24} \div \boxed{} = \underline{}$

13. $\dfrac{3}{36} \div \boxed{} = \underline{}$

14. $\dfrac{12}{15} \div \boxed{} = \underline{}$

15. $\dfrac{9}{12} \div \boxed{} = \underline{}$

16. $\dfrac{15}{18} \div \boxed{} = \underline{}$

17. $\dfrac{20}{30} \div \dfrac{10}{10} = \underline{}$

18. $\dfrac{25}{40} \div \boxed{1} = \underline{}$

19. $\dfrac{5}{50} \div \boxed{} = \underline{}$

20. $\dfrac{9}{15} \div \boxed{} = \underline{}$

21. $\dfrac{9}{18} \div \boxed{} = \underline{}$

22. $\dfrac{12}{36} \div \boxed{} = \underline{}$

23. $\dfrac{11}{22} \div \boxed{} = \underline{}$

24. $\dfrac{14}{35} \div \boxed{} = \underline{}$

Simplest Form

A fraction is in its simplest form when the greatest common factor (GCF) is 1.

$$\frac{6}{8} \div \frac{2}{2} = \frac{3}{4}$$

GCF of 6 and 8 is 2.
6 : 1, 2, 3, 6
8 : 1, 2, 4, 8

$$\frac{5}{8} \div \frac{1}{1} = \frac{5}{8} \quad \text{already simplified}$$

GCF of 5 and 8 is 1.
5 : 1, 5
8 : 1, 2, 4, 8

Simplify when necessary. If the fraction **does not** need simplifying, write "simplified."

1. $\frac{3}{5} =$ **simplified**

2. $\frac{15}{18} = \frac{5}{6}$

3. $\frac{3}{16} =$

4. $\frac{9}{18} =$

5. $\frac{4}{6} =$

6. $\frac{7}{8} =$

7. $\frac{15}{16} =$

8. $\frac{9}{12} =$

9. $\frac{7}{10} =$

10. $\frac{1}{8} =$

11. $\frac{3}{9} =$

12. $\frac{4}{5} =$

13. $\frac{16}{20} =$

14. $\frac{17}{18} =$

15. $\frac{3}{8} =$

16. $\frac{4}{9} =$

17. $\frac{9}{15} =$

18. $\frac{3}{4} =$

19. $\frac{6}{8} =$

20. $\frac{8}{8} =$

21. $\frac{8}{12} =$

22. $\frac{12}{16} =$

23. $\frac{5}{8} =$

24. $\frac{10}{12} =$

Think It Through

Sometimes you might not be able to think of the greatest common factor, but you can come up with a common factor.

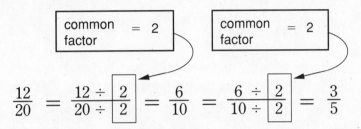

If you cannot think of the greatest common factor, keep dividing by common factors until the fraction is simplified to its lowest terms. But remember: finding the greatest common factor will help you simplify a problem quickly.

Simplify the fractions. Use more than one common factor if necessary.

1. $\dfrac{48}{60}$

2. $\dfrac{26}{78}$

3. $\dfrac{45}{60}$

4. $\dfrac{18}{30}$

5. $\dfrac{21}{63}$

6. $\dfrac{12}{30}$

7. $\dfrac{15}{30}$

8. $\dfrac{16}{32}$

9. $\dfrac{22}{44}$

10. $\dfrac{24}{36}$

11. $\dfrac{12}{18}$

12. $\dfrac{16}{40}$

Lowest Terms

Change each fraction to its simplest form. If the fraction **does not** need simplifying, write "simplified."

1. $\frac{16}{48} =$

2. $\frac{6}{24} =$

3. $\frac{9}{27} =$

4. $\frac{9}{36} =$

5. $\frac{18}{54} =$

6. $\frac{24}{36} =$

7. $\frac{9}{12} =$

8. $\frac{7}{14} =$

9. $\frac{16}{32} =$

10. $\frac{4}{12} =$

11. $\frac{3}{9} =$

12. $\frac{8}{24} =$

13. $\frac{15}{28} =$

14. $\frac{10}{12} =$

15. $\frac{21}{42} =$

16. $\frac{12}{16} =$

17. $\frac{14}{20} =$

18. $\frac{8}{16} =$

19. $\frac{6}{21} =$

20. $\frac{8}{18} =$

21. $\frac{15}{21} =$

22. $\frac{13}{24} =$

23. $\frac{15}{18} =$

24. $\frac{12}{14} =$

25. $\frac{15}{24} =$

26. $\frac{8}{12} =$

27. $\frac{10}{35} =$

28. $\frac{7}{20} =$

29. $\frac{15}{45} =$

30. $\frac{30}{42} =$

Changing Fractions to Decimals

Often it is necessary to change a fraction to a decimal. Think of the fraction bar as a division bar.

Change $\frac{1}{4}$ to a decimal.

Step 1: $\frac{1}{4}$ ←—— division bar   **Think:** $1 \div 4$

Step 2: $\frac{1}{4}$ $= 4\overline{)1.00}$

$$4\overline{)\begin{array}{l} .25 \\ 1.00 \end{array}}$$
$$\underline{8}$$
$$2\,0$$
$$\underline{2\,0}$$
$$0$$

so $\frac{1}{4} = .25$

Add as many zeros as you need until the division comes out exactly. The remainder will be zero.

Rewrite each fraction as a division problem. Then find its decimal value.

1. $\frac{1}{5}$ $5\overline{)1.0}$

2. $\frac{3}{4}$

3. $\frac{3}{8}$

4. $\frac{1}{8}$ $8\overline{)1.000}$

5. $\frac{3}{5}$

6. $\frac{7}{8}$

More Practice: Fractions to Decimals

Find the decimal value of these fractions. Remember to write as many zeros as necessary to have a remainder of zero.

1. $\dfrac{10}{8}$ = __.__ __

$$
\begin{array}{r}
1.2\,5 \\
8\,\overline{)\,1\,0.0\,0} \\
\underline{8} \\
2\,0 \\
\underline{1\,6} \\
4\,0 \\
\underline{4\,0} \\
0
\end{array}
$$

2. $\dfrac{18}{15}$ =

3. $\dfrac{9}{12}$ =

4. $\dfrac{1}{2}$ =

5. $\dfrac{5}{8}$ = .__ __ __

$$
8\,\overline{)\,5.0\,0\,0}
$$

6. $\dfrac{8}{10}$ =

7. $\dfrac{22}{16}$ =

8. $\dfrac{15}{4}$ =

9. $\dfrac{3}{12}$ = .__ __

$$
12\,\overline{)\,3}
$$

10. $\dfrac{24}{10}$ =

11. $\dfrac{14}{16}$ =

12. $\dfrac{11}{5}$ =

Working with Remainders

Some decimal forms of fractions have remainders after you have divided to the hundredths place.

- Divide to the hundredths place.
- Write all remainders as fractions: put the remainder over the original denominator.

$$\frac{1}{3} = 3\overline{)1.0\,0} \quad .33\tfrac{1}{3} \leftarrow \text{put the remainder over the denominator}$$

$$\begin{array}{r} .33\tfrac{1}{3} \\ 3\overline{)1.0\,0} \\ \underline{9} \\ 1\,0 \\ \underline{9} \\ 1 \end{array} \leftarrow \text{remainder}$$

$$\frac{5}{8} = 8\overline{)5.0\,0} \qquad .62\tfrac{4}{8} = .62\tfrac{1}{2} \leftarrow \text{simplify the remainder}$$

$$\begin{array}{r} .62\tfrac{4}{8} \\ 8\overline{)5.0\,0} \\ \underline{4\,8} \\ 2\,0 \\ \underline{1\,6} \\ 4 \end{array} \leftarrow \text{remainder}$$

Find the decimal form of each fraction. Divide to the hundredths place and write all remainders as fractions.

1. $\frac{2}{3}$

$$\begin{array}{r} .6 \\ 3\overline{)2.0\,0} \\ \underline{1\,8} \\ 2\,0 \end{array}$$

4. $\frac{1}{9}$

2. $\frac{5}{6}$ $\qquad 6\overline{)5.0\,0}$

5. $\frac{2}{9}$

3. $\frac{7}{8}$

6. $\frac{4}{9}$

Practice Helps

- Carry out all decimals to the hundredths place (two decimal places).
- Change all remainders to fractions.

$$\frac{10}{7} = 7\overline{\smash{)}\,10.00} \quad 1.42\tfrac{6}{7}$$

$$\begin{array}{r} 1.42\ \tfrac{6}{7} \\ 7\,)\,\overline{10.00} \\ \underline{7} \\ 30 \\ \underline{28} \\ 20 \\ \underline{14} \\ 6 \end{array}$$

← put the remainder over the denominator

← remainder

Find the decimal form of each fraction. Divide to the hundedths place and write all remainders as fractions.

1. $\frac{8}{18}$

$$\begin{array}{r} .4 \\ 18\,)\,\overline{8.00} \\ \underline{72} \\ -\ - \\ =\ = \\ - \end{array}$$

finish

4. $\frac{3}{27}$

2. $\frac{15}{7}$

5. $\frac{33}{21}$

3. $\frac{6}{7}$

6. $\frac{15}{9}$

50

Half-Inch Measurements

Find the measurement in inches for each line.

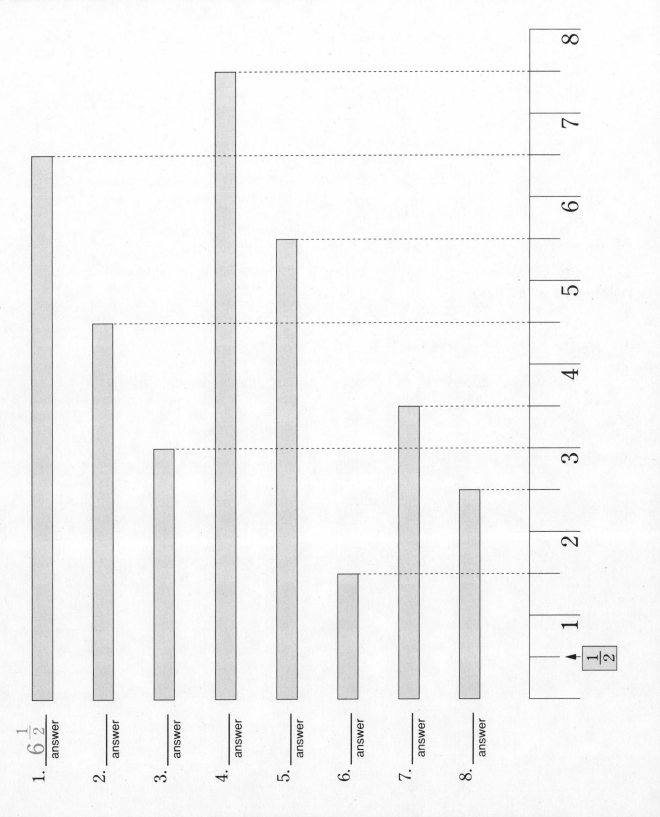

1. $6\frac{1}{2}$
   <u>answer</u>

2. ___
   <u>answer</u>

3. ___
   <u>answer</u>

4. ___
   <u>answer</u>

5. ___
   <u>answer</u>

6. ___
   <u>answer</u>

7. ___
   <u>answer</u>

8. ___
   <u>answer</u>

Quarter-Inch Measurements

Find the measurement in inches for each line.

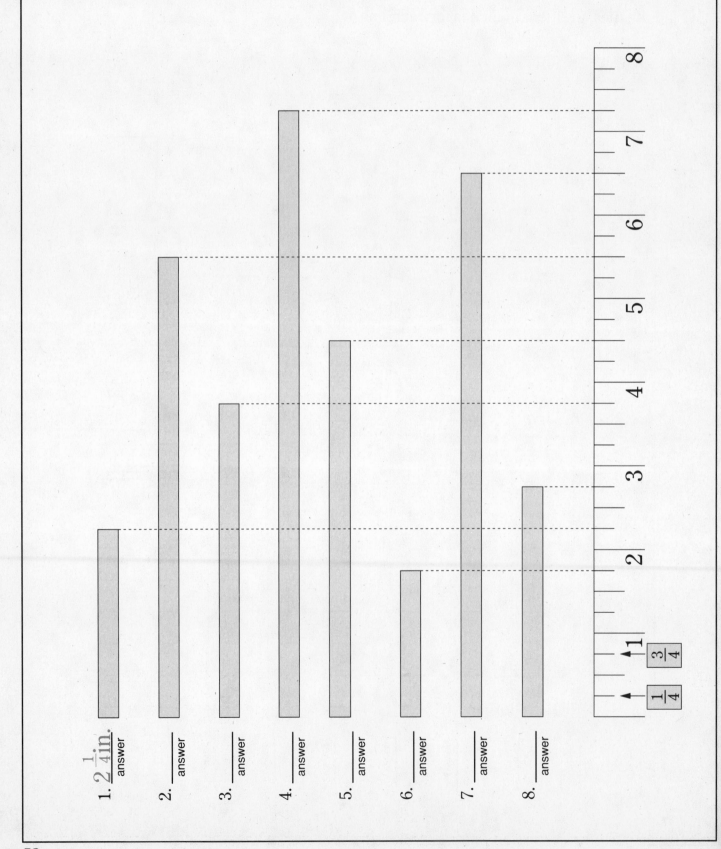

1. $2\frac{1}{4}$in.
   <u>answer</u>

2. _____
   answer

3. _____
   answer

4. _____
   answer

5. _____
   answer

6. _____
   answer

7. _____
   answer

8. _____
   answer

$\frac{3}{4}$

$\frac{1}{4}$

Eighth-Inch Measurements

Find the measurement in inches for each line.

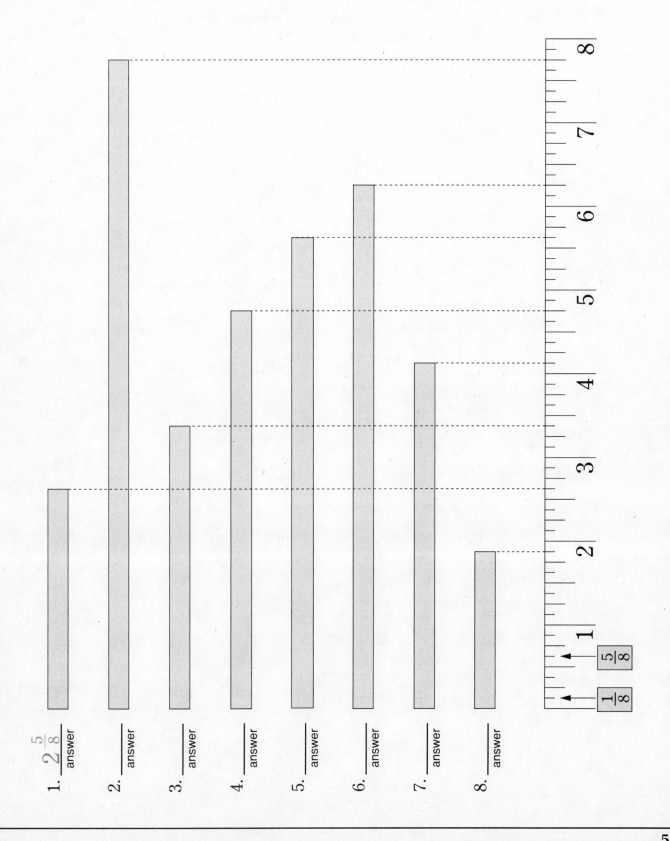

1. $2\frac{5}{8}$
   answer

2. _____
   answer

3. _____
   answer

4. _____
   answer

5. _____
   answer

6. _____
   answer

7. _____
   answer

8. _____
   answer

Draw the Measurements

Draw lines for the following lengths.

1. $5\frac{1}{4}$ inches
2. $1\frac{1}{2}$ inches
3. $4\frac{3}{4}$ inches

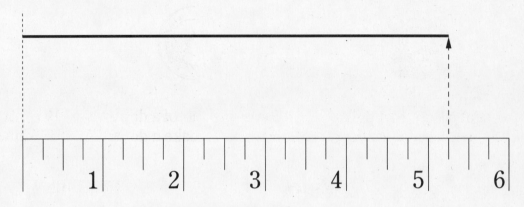

4. $3\frac{1}{4}$ inches
5. $5\frac{3}{4}$ inches
6. $1\frac{5}{8}$ inches

7. $5\frac{7}{8}$ inches
8. $2\frac{3}{8}$ inches
9. $4\frac{9}{16}$ inches

10. $4\frac{5}{16}$ inches
11. $1\frac{1}{4}$ inches
12. $3\frac{1}{2}$ inches

Money as Fractions

Penny

1 cent = $\frac{1}{100}$ of a dollar

100 pennies make a dollar

Nickel

5 cents = $\frac{1}{20}$ of a dollar

20 nickels make a dollar

Dime

10 cents = $\frac{1}{10}$ of a dollar

10 dimes make a dollar

Quarter

25 cents = $\frac{1}{4}$ of a dollar

4 quarters make a dollar

Half-Dollar

50 cents = $\frac{1}{2}$ of a dollar

2 half-dollars make a dollar

Find the fraction of a dollar. Simplify if necessary.

1. 4 dimes = $\frac{4}{10}$ or $\frac{2}{5}$ of a dollar

6. 25 pennies = $\frac{\square}{\square}$ or $\frac{\square}{\square}$ of a dollar

2. 10 pennies = $\frac{10}{100}$ or $\frac{\square}{\square}$ of a dollar

7. 10 nickels = $\frac{\square}{\square}$ or $\frac{\square}{\square}$ of a dollar

3. 5 nickels = $\frac{\square}{20}$ or $\frac{\square}{\square}$ of a dollar

8. 7 nickels = $\frac{\square}{\square}$ of a dollar

4. 3 quarters = $\frac{\square}{\square}$ of a dollar

9. 1 half-dollar = $\frac{\square}{\square}$ of a dollar

5. 5 dimes = $\frac{\square}{\square}$ or $\frac{\square}{\square}$ of a dollar

10. 8 dimes = $\frac{\square}{\square}$ or $\frac{\square}{\square}$ of a dollar

What Coin Is Missing?

Find the missing coin that completes each fraction of a dollar.

1.

   $\frac{1}{4}$ of 1 dollar

2.

   $\frac{3}{10}$ of 1 dollar

   (Hint: 25 cents + 5 cents = $\frac{30}{100}$ or $\frac{3}{10}$)

3.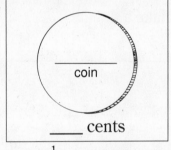

   $\frac{1}{2}$ of 1 dollar

4.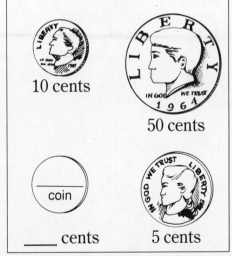

   $\frac{3}{4}$ of 1 dollar

5.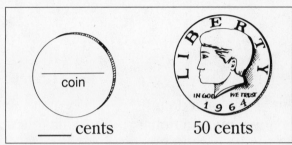

   $\frac{3}{4}$ of 1 dollar

6.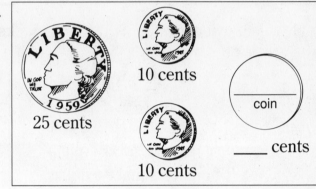

   $\frac{1}{2}$ of 1 dollar

7.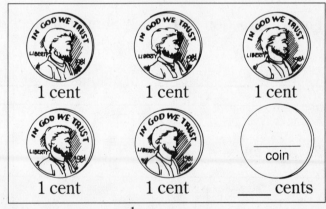

   $\frac{1}{10}$ of 1 dollar

8.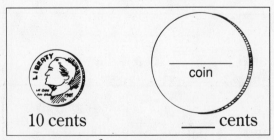

   $\frac{6}{10}$ of 1 dollar

Fill Up the Gallons

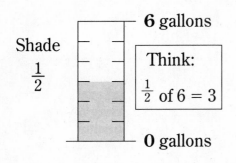

Shade $\frac{1}{2}$

6 gallons

Think:
$\frac{1}{2}$ of 6 = 3

0 gallons

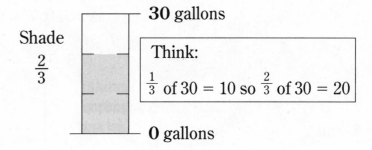

Shade $\frac{2}{3}$

30 gallons

Think:
$\frac{1}{3}$ of 30 = 10 so $\frac{2}{3}$ of 30 = 20

0 gallons

A. $\frac{1}{2}$ of 6 gallons = ____ gallons
_{fill in}

B. $\frac{2}{3}$ of 30 gallons = ____ gallons
_{fill in}

Shade the drawings and then fill in the blanks.

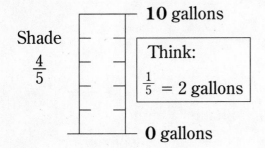

Shade $\frac{4}{5}$

10 gallons

Think:
$\frac{1}{5}$ = 2 gallons

0 gallons

1. $\frac{4}{5}$ of 10 gallons = ____ gallons

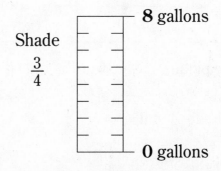

Shade $\frac{3}{4}$

8 gallons

0 gallons

3. $\frac{3}{4}$ of 8 gallons = ____ gallons

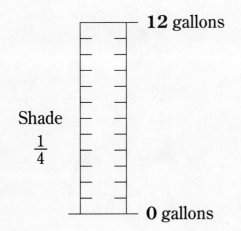

Shade $\frac{1}{4}$

12 gallons

0 gallons

2. $\frac{1}{4}$ of 12 gallons = ____ gallons

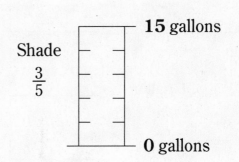

Shade $\frac{3}{5}$

15 gallons

0 gallons

4. $\frac{3}{5}$ of 15 gallons = ____ gallons

Find the Measurements

1. Show $\frac{1}{4}$ full

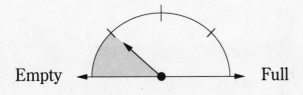

2. Show $\frac{3}{4}$ full

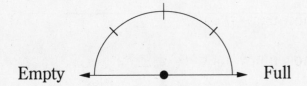

3. Show $\frac{5}{8}$ full

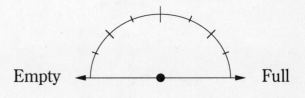

4. Show $\frac{7}{8}$ full

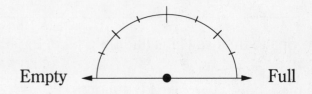

Use the pictures to answer the questions.

5. Shade $\frac{2}{3}$ full

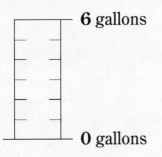

6. $\frac{1}{3}$ of 6 gallons = _____

7. $\frac{2}{3}$ of 6 gallons = _____

8. $\frac{3}{3}$ of 6 gallons = _____

9. $\frac{1}{2}$ of 6 gallons = _____

10. Shade $\frac{3}{4}$ full

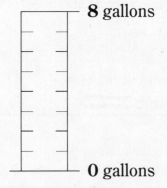

11. $\frac{1}{4}$ of 8 gallons = _____

12. $\frac{1}{2}$ of 8 gallons = _____

13. $\frac{3}{4}$ of 8 gallons = _____

14. $\frac{4}{4}$ of 8 gallons = _____

Real-Life Applications

To find one part of a total, divide the denominator into the total.

$\frac{1}{4}$ of an hour = $\frac{1}{4}$ of 60 minutes

$$\begin{array}{r} 1\ 5 \text{ minutes} \\ 4 \overline{)\ 6\ 0} \text{ minutes} \\ \underline{4} \\ 2\ 0 \\ \underline{2\ 0} \\ 0 \end{array}$$

**Fractions
of an Hour**

1 hour = 60 minutes

1. $\frac{1}{2}$ of an hour is how many minutes? ____

2. $\frac{1}{6}$ of an hour is how many minutes? ____

3. $\frac{1}{3}$ of an hour is how many minutes? ____

4. $1\frac{1}{4}$ hours is how many minutes? ____

    60   +   ?
  minutes minutes

**Fractions
of a Year**

1 year = 12 months

5. $\frac{1}{2}$ of a year (semi-annual) is how many months? ____

6. $\frac{1}{4}$ of a year (quarterly) is how many months? ____

7. $\frac{1}{3}$ of a year is how many months? ____

8. $1\frac{1}{2}$ years is how many months? ____

  12   +   ?
months months

**Fractions
of a Dollar**

1 dollar = 100 cents

9. $\frac{1}{2}$ of a dollar is how many cents? ____

10. $\frac{1}{4}$ of a dollar is how many cents? ____

11. $\frac{3}{4}$ of a dollar is how many cents? ____
(Hint: use your answer to question 10 as a starting point.)

12. $\frac{1}{10}$ of a dollar is how many cents? ____

Review

1. $\frac{1}{6}$ of an hour is how many minutes? ____

2. $1\frac{1}{3}$ years are how many months? ____

3. Geneva wants only $\frac{1}{3}$ as much orange punch as the recipe calls for. How much would the new recipe take?

New Recipe	Original Recipe
a) ____	9 oranges
b) ____	3 gallons of ice cream
c) ____	12 cups of ginger ale

4. What fraction of a dollar is shown? ____

quarter

dime

nickel

dime

Simplify the fractions.

5. $\frac{8}{12} =$

6. $\frac{10}{12} =$

7. $\frac{4}{28} =$

8. $\frac{6}{15} =$

9. $\frac{32}{40} =$

10. $\frac{15}{25} =$

11. $\frac{18}{36} =$

12. $\frac{16}{30} =$

Change each fraction to a decimal.

13. $\frac{7}{8}$

14. $\frac{12}{15}$

15. $\frac{5}{4}$